YOUR KNOWLEDGE HAS VALUE

- We will publish your bachelor's and
  master's thesis, essays and papers

- Your own eBook and book -
  sold worldwide in all relevant shops

- Earn money with each sale

Upload your text at www.GRIN.com
and publish for free

Water and Cities in the Face of Future Challenges. An Examination of Adaption Measures for Urban Areas

Juri Felde

Bibliographic information published by the German National Library:

The German National Library lists this publication in the National Bibliography; detailed bibliographic data are available on the Internet at http://dnb.dnb.de.

ISBN: 9783346848376
This book is also available as an ebook.

University of Osnabrück

Faculty 1: Cultural and Social Studies

Department of Geography

Course Title: **Water: Risk and source of life – an inter- and transdisciplinary perspective**

WS 2022/2023

Water and Cities in the Face of Future Challenges:

An Examination of Adaptation Measures for Urban Areas

Juri Felde

Master of Education: Geography / English and American Studies

Table of Contents

1. Introduction

Global challenges of the 21st century can cause a significant threat to water security in urban areas in the future. Rising temperatures, increased precipitation, and more frequent extreme weather events are transforming the availability and distribution of water resources, leading to increased pressure on water systems and infrastructure. Furthermore, the reduction of permeable green space and loss of floodable urban spaces will increase the vulnerability of urban environments (O'Donnell et al. 2020, Romano & Akhmouch 2019).

Cities are already economic engines of entire regions. About 600 urban regions on earth with 20% of the world's population contribute 60% of the GWP (gross world product). Moreover, By the year 2050, 70 % of the world's population will be living in cities and urban areas. Therefore, the demand for water will increase significantly as well as the number of water stressed areas. Despite the fact that the number of people who have no access to reliable water resources will grow significantly, some areas will be lacking access to any water sources at all or dealing with water pollution issues. (Koop & Leuween 2017, Romano & Akhmouch 2019).

As cities around the globe continue to grow and urbanize, the need for effective adaptation measures to address these challenges becomes increasingly urgent. This essay attempts to identify the known vulnerabilities of urban environments and then aims to examine the various opportunities for adaptation in order to ensure water security in the face of global challenges of the 21st century and especially water related extreme weather events.

Through a literature review, this paper aims to explore a concise outlook on strategies cities can employ in order to build resilience and complete a transition to urban areas fit for the future.

2. Future Trends and Challenges in Urban Water Resource Management and Infrastructure

As the planet warms, the frequency and intensity of extreme weather events grows. This means that the impact of flooding and heavy rainfall events (too much water), the impact of drought and scarcity (not enough water) as well as the pollution of water must be calculated to a new extent. Especially urban landscapes, which are characterized by a strong increase in population and thus by an increased vulnerability to such events, are affected by this. (Romano & Akhmouch 2019). This chapter aims to identify the main factors that lead to an increased vulnerability in cities.

A study from Great Britain (O' Donnel et al. 2020) serves as example. Since urban environments are areas with large impermeable surfaces, one main threat consists of intra-urban runoff during heavy precipitation events. The sealing of urban spaces increased by 22 % compared to the year 2001. Hence, the study emphasises that urbanization could increase the damage caused by urban surface flooding by 60 to 220 %. As a result, sewerage and wastewater systems could reach the limits of their capacity. This could lead to a widespread drinking water pollution and an utilisation of sanitation systems. In addition, water supply infrastructure in cities can be extremely old, making it difficult and costly to repair this infrastructure after it has been damaged. (O' Donnel et al. 2020).

Another significant physical factor is the loss of floodable urban spaces. Former green spaces and brownfields give way to infrastructure and buildings which to a decreased capacity of absorbing rainfall in urban areas. Furthermore, this leads to changing water pathways during precipitation events that 'surcharge the urban drainage system' (O' Donnel et al 2020: 18). Additionally, cities located in coastal areas are exposed to flood risk resulting from storm surges and of course sea level rise due to climate change.

2.1 Social and Economic Factors Shaping the Future of Urban Water Management

Locations with water views or access to water have recently become much more attractive. It is not only the harmonious view that makes these locations so attractive, but above all their profitability. Sites on the waterfront are expensive and usually sold out quickly. Those who can afford to build at water sites, whether residential or commercial, can expect a profitable investment in the real estate market. The key factors are proximity to water and an urban society that can afford the costs.

As cities are currently subject to strong dynamic construction processes and the speed of structural change in an urban environment could accelerate in the future, the construction of demanding and expensive buildings is taking place. If urban land use planning does not intervene in this process in a sustainable manner with appropriate development strategies, this can lead to poor urban planning and thus to increased vulnerability of new and expensive buildings (O' Donnel et al. 2020). Since many interests and conflicts of use have to be reconciled in the process of urban planning, this process is a particular challenge for the future. There is a need for multiple points of view from different stakeholder perspectives (city population, economists, town planners, experts, political leaders etc.) to consolidate different interests.

In addition, minimizing the risk of flood disasters is a major future challenge for urban planners considering the rapid change of climatic conditions around the globe (Barroca et al. 2006).

3. The Impact of Hurricane 'Sandy' in New York (2012)

In this chapter, this essay aims to take a closer look at the impact of Hurricane Sandy in 2012, which caused an overall damage of 19 billion dollars. What makes 'Sandy' so special is that hurricanes of this magnitude have a one percent annual chance of occurring on the Northeast Coast of the United States. Normally, hurricanes are unleashed on the southeast coast of the U.S., and it is very rare for a hurricane of this magnitude to spread further north. Due to the effects of climate change, it is believed that such events will occur more frequently in the future in areas where they were not expected (Cimellaro et al. 2019).

Although financial resources were made available very quickly and the people responsible began reconstruction work immediately after the disaster, many of the projects initiated are still far from completion a decade after the disaster. The planning of the reconstruction turns out to be particularly difficult as important infrastructure lies in the 100-year floodplain areas. The office of the New York City comptroller published following figures (Lander 2022):

- 79% of transportation and utility land uses that support our electric and gas utilities, rail yards, airports, docks and piers, bridges, tunnels, and highways.

- 67% of open space and outdoor recreation areas, from neighborhood parks that provide vital space for local residents to iconic parks that draw visitors.
- 46% of the city's industrial and manufacturing that house waste transfer stations, construction businesses, warehouses and distribution centers – industries on which our local economy and supply chain rely

These numbers illustrate how difficult it is for all stakeholders to set reconstruction priorities and to plan for sustainable urban infrastructure that decreases the vulnerability of urban environments to future extreme weather events. Despite the ongoing reconstruction measures, Lander (2022) has noted that 'market rate values of real estate in the 100-year floodplain have increased to over $176 billion – a 44% increase since Superstorm Sandy'.

Therefore, it becomes clear how much time is needed to replan urban structures that have been built up over decades and to adapt them to future challenges. It also reveals how important it is to involve a wide range of stakeholders in planning processes that lead to reduced vulnerability. Thus, targeted spatial planning and zoning that reduces vulnerability and remains open for new adaptions is crucial for future urban planning.

4. Adaption Measures for a Rapidly Changing Future

In the past, and even today, societies have not been afraid to settle in floodplains. What differs between past and present, however, is the change in the perception of flood risk that societies face through decisions to settle in floodable areas. In the past, people increasingly protected potential floodplains from flooding by building dams. This led to a false perception of risk because people lived and farmed near dams with the notion of being safe. If an extreme weather event does occur, leading to a levee breach, for example, the potential damage is enormous due to the distorted perception of risk- a phenomenon also known as the Levee effect (Di Baldassarre et al. 2013).

Today, the focus is shifting from the idea of designing an environment that is as safe as possible to the idea of risk reduction and disaster resilience to minimize the negative impacts of extreme weather related events. In order to determine the state, development and evaluation of disaster resilience of an urban environment, appropriate scientific tools must be developed to combine different findings from different research fields into measurable results (Cai et al. 2018).

In addition, to express it in the words of Romano & Akhmouch (2019: 2) 'often water crises are water governance crises: managing water risks of too much, too little, and too polluted water is all the more challenging if the roles and responsibilities are not clearly allocated, stakeholders are not engaged, information is not shared and the capacities are not adequate to anticipate and tackle the risks`.

A framework to conquer those issues is presented by the OECD (The Organisation for Economic Co-operation and Development) as Principles of Water Governance (*Figure 1*).

Figure 1: The Principles of Water Governance by the OECD (Romano & Akhmouch 2019: 3)

This framework aims to prevent conditions such as poor governance of water resources, an aging infrastructure, and increased hazards as a result of climate change. It also should help representatives to improve water policy and therefore to mitigate the risk of flood, install instruments to prepare for drought and prevent water resources from pollution.

According to this OECD framework, water governance should be effective, so water actors have clear policy goals and are well trained and most importantly work together. To make water governance efficient data should be produced in a timely manner and funds must be mobilized to meet societies needs. Also, frameworks for innovation and action should be encouraged. The 'Trust & Engagement' segment means that a participatory decision-making should be supported by including communities, individuals and all relevant stakeholders to promote integrity and transparency. In addition to the theoretical framework of the OECD, which intends to help policy makers better understand the complexities of water governance and make informed decisions, the following section will focus on specific structural measures that can make an urban environment more resilient to water-related extreme weather events.

Radhakrishnan et al. (2019) introduce the concept of Water Sensitive Cities and suggest that when planning adaption measures, flexibility should be considered as an important factor to changing circumstances. It is necessary to identify factors that reduce negative impacts and increase positive ones. In order to do that, the authors introduce the concept of a flexible water sensible adaption response. This concept aims to help urban planners to identify where the water sensible adaption should be implemented best. Furthermore, it helps to decide what type of adaptions are contributing the most to the concept of a water sensible city. By establishing diverse adaptation concepts, it is possible to retain a degree of flexibility for future planning and thus uncertain factors can be taken into account to a limited extent. The concept intends to go beyond water sensitive architecture by reinforcing water sensitive behaviours.

In order 'to increase resilience towards flooding in designing and planning systems for water sensitivity' (Radhakrishnan et al. 2019: 1) a combination of different strategy types is suggested (*Figure 2*).

Figure 2: 4-RAP Model of strategies. 4R= retain, relive, resist, retreat, A= accomodate, P= prepare (Radhakrishnan et al. 2019: 2)

This model offers a wide range of adaption measures. Among other things, some crucial measures are presented in this model. For example, the creation of more green spaces to prevent the progressive sealing of surfaces. Furthermore, the establishment of natural retention spaces in urban areas, building a resilient water infrastructure that should remain operational even in the event of a disaster. Additionally, there should be a new distribution of land use zoning as well as the provision of improved risk maps that not only include areas endangered by flood but also locations of critical infrastructure. It is also necessary to implement case- specific instruction in case of an emergency (Radhakrishnan 2019).

5. Conclusion

The goal of this essay was to identify future water-related threats to urban environments. In the context of the global challenges of the 21st century, adaptation measures were investigated based on a literature review. It is detected that a paradigm shift is taking place in the area of risk perception. It turns out that the idea of living in perfect safety is slowly being replaced by the idea of being prepared for as many events as possible. Designing an urban area by keeping the negative effects of floods and heavy rainfalls to a minimum is crucial. In order to prepare cities for upcoming challenges, it is important to involve as many relevant stakeholders as possible in the planning process.

The OECD proposes the water governance framework for this purpose (Romano & Akhmouch 2019), while Radhakrishnan et al. 2019 suggest an implementation of diverse technical flood protection measures. Modelling flood probability scenarios and simulating future conditions in computer models will definitely play a big role in disaster risk measurement and reduction in the future (Cai et al . 2018, Tsakiris 2014).

Although technical solutions will be an important part of the solve, it is important to avoid separating water from its social context and therefore to apply a more hydrosocial approach regarding the society-water relationship in the sense of Linton & Buds (2012). Therefore, it is crucial to understand the complicated network of human-water relationships in as many aspects as possible. Other issues, such as challenges cities may face in implementing the introduced frameworks as well as more recent case studies, unfortunately cannot be explored within the scope of this essay.

Yet, a brief look at the consequences of hurricane Sandy for the city of New York showed that a sustainable restructuring of a city takes several decades. Nevertheless, there is one important conclusion. The overall focus of water security in urban areas is on integrating a wide variety of approaches in order to be able to adapt as flexible as possible to rapidly changing and partly unknown conditions in the future.

Bibliography

Barroca, B., Bernardara, P., Mouchel, J. M., and Hubert, G (2006): Indicators for identification of urban flooding vulnerability, Nat. Hazards Earth Syst. Sci., 6, 553–561, https://doi.org/10.5194/nhess-6-553-2006. https://nhess.copernicus.org/articles/6/553/2006/nhess-6-553-2006.pdf   (25.01.2023)

Cai, H.; Lam, N. S.N.; Qiang, Y.; Zou, L.; Correll, R. M.; Mihunov, V. (2018): *A Synthesis of Disaster Resilience Measurement Methods and Indices. International Journal of Disaster Risk Reduction,* doi:10.1016/j.ijdrr.2018.07.015

Cimellaro, G. P.; Crupi,P.; Kim, H.U.; Agrawal, A (2019): Modeling interdependencies of critical infrastructures after hurricane Sandy, International Journal of Disaster Risk Reduction, doi: 10.1016/j.ijdrr.2019.101191.

Di Baldassarre, G.; Viglione, A.; Carr, G.; Kuil, L.; Salinas, J. L.; Blöschl, G. (2013): *Socio-hydrology: conceptualising human-flood interactions. Hydrology and Earth System Sciences, 17(8), 3295–3303.* doi: 10.5194/hess-17-3295-2013

Koop, S. H. A.; van Leeuwen, C. J. (2017): The challenges of water, waste and climate change in cities. Environment, Development and Sustainability, 9(2), 385–418. doi:10.1007/s10668-016-9760-4 https://link.springer.com/content/pdf/10.1007/s10668-016-9760- 4.pdf?pdf=button (21.01.2023)

Lander, B. (2022): Ten Years After Sandy – Barriers To Resilience. New York City Comptroller. https://comptroller.nyc.gov/reports/ten-years-after-sandy (20.02.2023)

Linton, J., Budds, J. (2013): The hydrosocial cycle: Defining and mobilizing a   relational-dialectical   approach   to water. Geoforum. https://www.researchgate.net/publication/259089822_The_Hydrosocial_ Cycle_Defining_and_Mobilizing_a_Relationa Dialectical_Approach_to_Water (22.01.2023).

O'Donnell, E.C.; Thorne, C. R. (2020): *Drivers of future urban flood risk.   Philosophical   Transactions   of the Royal Society A: Mathematical, Physical and Engineering   Sciences,   378(2168),   20190216–.* doi:10.1098/rsta.2019.0216 https://royalsocietypublishing.org/doi/10.1098/rsta.2019.0216 (26.01.2023).

Radhakrishnan, M.; Pathirana, A.; Ashley, R.M.; Gersonius, B.; Zevenbergen, C. (2018): Flexible adaptation planning for water sensitive cities. Cities, doi:10.1016/j.cities.2018.01.022

Romano, O.; Akhmouch, A. (2019): Water Governance in Cities: Current Trends and Future Challenges. Water, 11(3), doi:10.3390/w11030500

Tsakiris, G (2014): Flood risk assessment: concepts, modelling, applications. Natural Hazards and Earth System Sciences, 14(5), 1361–1369. doi:10.5194/nhess-14-1361-2014